THE POETRY OF POLONIUM

The Poetry of Polonium

Walter the Educator

Silent King Books

SILENT KING BOOKS

SKB

Copyright © 2024 by Walter the Educator

All rights reserved. No part of this book may be reproduced in any manner whatsoever without written permission except in the case of brief quotations embodied in critical articles and reviews.

First Printing, 2024

Disclaimer
This book is a literary work; poems are not about specific persons, locations, situations, and/or circumstances unless mentioned in a historical context. This book is for entertainment and informational purposes only. The author and publisher offer this information without warranties expressed or implied. No matter the grounds, neither the author nor the publisher will be accountable for any losses, injuries, or other damages caused by the reader's use of this book. The use of this book acknowledges an understanding and acceptance of this disclaimer.

"Earning a degree in chemistry changed my life!"
- Walter the Educator

dedicated to all the chemistry lovers, like myself, across the world

POLONIUM

There lies a treasure, a subtle trance.

POLONIUM

Polonium, rare and enigmatic,

POLONIUM

In its essence, secrets enigmatic.

POLONIUM

From Marie Curie's deft embrace,

POLONIUM

Sprung forth this element's mystic grace.

POLONIUM

A silent allure, a deadly charm,

POLONIUM

In its nucleus, a radiant swarm.

POLONIUM

In shadows deep, it quietly dwells,

POLONIUM

A clandestine tale that history tells.

POLONIUM

With atomic weight, and half-life short,

POLONIUM

It whispers secrets of nuclear sorts.

POLONIUM

A glimmering poison, a clandestine sprite,

POLONIUM

In the heart of darkness, it shines so bright.

POLONIUM

With alpha decay, it sheds its cloak,

POLONIUM

Releasing energy with every stroke.

POLONIUM

In clandestine labs, where minds conspire,

POLONIUM

Polonium's whispers fuel desire.

POLONIUM

A clandestine kiss, a deadly bloom,

POLONIUM

In its embrace, there lies the gloom.

POLONIUM

From radium's womb, it takes its cue,

POLONIUM

A progeny born of atoms few.

POLONIUM

In leaden shells, it finds its home,

POLONIUM

A silent sentinel, free to roam.

POLONIUM

Yet in its shadow, lies a tale,

POLONIUM

Of lives cut short, of ships set sail.

POLONIUM

From espionage to tragic end,

POLONIUM

Polonium's touch, it does portend.

POLONIUM

But in the depths, beyond the gloom,

POLONIUM

There lies a hope, a distant bloom.

POLONIUM

For in the study of this element rare,

POLONIUM

Lies the key to secrets, beyond compare.

POLONIUM

In quantum fields, where particles play,

POLONIUM

Polonium's dance leads the way.

POLONIUM

A symphony of chaos, a cosmic song,

POLONIUM

In its atoms, the universe belongs.

POLONIUM

So let us ponder, with minds aglow,

POLONIUM

The mysteries of Polonium's glow.

POLONIUM

For in its essence, we may find,

POLONIUM

The secrets of the universe entwined.

POLONIUM

In labs of science, and halls of lore,

POLONIUM

Polonium's tale, forevermore.

POLONIUM

A silent witness, to the grand design,

POLONIUM

In its atoms, the universe aligns.

POLONIUM

ABOUT THE CREATOR

Walter the Educator is one of the pseudonyms for Walter Anderson. Formally educated in Chemistry, Business, and Education, he is an educator, an author, a diverse entrepreneur, and he is the son of a disabled war veteran. "Walter the Educator" shares his time between educating and creating. He holds interests and owns several creative projects that entertain, enlighten, enhance, and educate, hoping to inspire and motivate you.

Follow, find new works, and stay up to date
with Walter the Educator™
at WaltertheEducator.com

www.ingramcontent.com/pod-product-compliance
Lightning Source LLC
La Vergne TN
LVHW010620070526
838199LV00063BA/5216